YOUR KNOWLEDGE HAS VALUE

- We will publish your bachelor's and master's thesis, essays and papers

- Your own eBook and book -
sold worldwide in all relevant shops

- Earn money with each sale

Upload your text at www.GRIN.com
and publish for free

Data Warehousing of Solid Minerals in Nigeria. A Study of South South Region Nigeria

Udeme Usanga

Bibliographic information published by the German National Library:

The German National Library lists this publication in the National Bibliography; detailed bibliographic data are available on the Internet at http://dnb.dnb.de.

ISBN: 9783346933522
This book is also available as an ebook.

© GRIN Publishing GmbH
Trappentreustraße 1
80339 München

All rights reserved

Print and binding: Books on Demand GmbH, Norderstedt, Germany
Printed on acid-free paper from responsible sources.

The present work has been carefully prepared. Nevertheless, authors and publishers do not incur liability for the correctness of information, notes, links and advice as well as any printing errors.

GRIN web shop: https://www.grin.com/document/1387590

Data Warehousing of Solid Minerals in Nigeria: A Study of South South Region Nigeria

Geology Department, University of Calabar, Calabar, Cross River State, Nigeria

**Agricultural Extensnion and Management
Computer Science Department
NRCRI- Federal College of Agriculture, Ishiagu, Ebonyi State, Nigeria**

In recent years, decision support systems otherwise called Data Warehouse have become an integral part of organization's decision making strategy. Organizations thessdays are competing in the globalmarket. In order for any organization to gain competitive advantage over the others and also to help make better decisions, Data warehousing cum Data Mining are now playing a significant role in strategic decision making, nationally and globally. It helps in better decisions, streamline work-flows, and provide better customer services. This paper gives the report about developing data warehouse for solid minerals in Nigeria, using a model of management system as a case study to capture solid minerals in the South South Region of Nigeria. It describes the process of data warehouse design and development using Microsoft SQL Server Analysis Services. It also outlines the development of a data cube as well as application of Online Analytical processing (OLAP) tools and Data mining tools in data analysis. It was concluded that the effective use of Data-Warehousing and Datamining in solid minerals resource management system will promote the rapid growth of the Nigerian economy and offer information on past, present and future prosects of mineral deposits, identification, exploitation, exploration and utilization for economic development.

Introduction

Nigeria lies approximately between Latitudes 4°N and 15°N of the Equator and Longitudes 3°E and 14°E, of the Greenwich Meridan, within the Pan African mobile belt in between the West African and Congo cratonsn (NEITI, 2017; Damilek, 2017). The Geology of Nigeria is dominated by three major rock types: the crystalline (igneous and metamorphic rocks) and sedimentary rocks both occurring approximately in equal proportions. The crystalline igneous and metamorphic rocks constitute the Precambrian- Palaeozoic basement complex, which occur in the Eastern Region of the country and extend through the North Central to the North Eastern part of Nigeria. The Sedimentary Basins comprises of seven (7) inland basins namely: the Niger Delta, the Anambra Basin, the Benue Trough, the Chad Basin, the Sokoto Basin, the Bida-Nupe Basin and the Dahomey Basin, all infill with sediments varying in age from the Cretaceous era to recent times (Bamali, 2011; Olade, 2019).

Nigeria's Solid Minerals Sector

The Nigerian economy is ranked as the largest in Africa and the 30th in the world, with a Gross Domestic Product (GDP) of approximately \$400 billion (US dollars) in 2018 (NEITI, 2018; Olade, 2019). There are five dominant sectors: (1) Agriculture, (2) Oil and gas (petroleum), (3) Mining (solid minerals), (4) Manufacturing and (5) Services. Nigeria is a highly-populated emerging market with a rapidly growing manufacturing, and services sectors. Unfortunately, the economy is still very much monolithic and dependent on the oil and gas sector, which contributes about 40% of the nominal GDP and over 90% of export earnings and 75% of gross revenues (Bamali, 2011; Damilek, 2017). Petroleum export is also the country's major earner of foreign exchange. As revenue from crude oil has declined sharply in recent years, Nigeria has struggled to find other revenue sources by trying to diversify its economy through energizing the other sectors of the economy. Agriculture, currently contributes about 24% of the GDP, while contribution from manufacturing has fluctuated and recently stabilized at about 21% (MMSD, 2016).

Mining activities began in 1902 and the earliest recorded mineral mined was tin ore (MMSD, 2017). In 2015, Nigeria's solid minerals and mining sector accounted for a paltry 0.3% of the GDP which increased slightly to about 0.5% in 2018. The sector has contributed only 0.02% of Nigeria's total exports. The sector's performance can be compared to contributions in other African countries to their GDP as follows: Botswana (38%), Democratic Republic of Congo (25%), South Africa (18%), Zambia (18%), Guinea (12%) and Ghana (7%) (MMSD, 2017). There has been some progress in Nigeria's solid mineral sector in the past few years due to the implementation of new policies, but the change is still miniscule. Between 2010 and 2015, reports indicate that the contribution of the solid minerals sector to the GDP doubled from $143 million to $290 million and to about $2 billion in 2018 (NEITI, 2018; n). The increase in revenues can be attributed not to higher mineral production as such, but mostly to more efficient revenue collection mechanisms and the increased foreign investment through the purchase of mineral titles.

The relatively low production output of solid minerals in Nigeria constitutes a major challenge to the growth of the states' and the national economy, particularly because this key sector offers such a great potential in the diversification of the Nigerian economy. The current situation was not always the case. Prior to the 1960s, mining was a primary mainstay of the economy, contributing up to 50% of the GDP. This high level of contribution declined precipitously, starting in the 1960s after the discovery of fossil fuel and gas which diverted the interest of government and investors from agriculture and mining of solid minerals. Moreover, the 30 month civil war of 1967 to 1970s and the attendant political instability and territorial insecurity led to the mass exodus of foreign mining companies and their experts, and migration of some indigenous miners, which led to the eventual abandonment of mining operations in some pats of the country. By the 1970s, the contribution of solid minerals to the economy of Nigeria had fallen to less than 5%, with further decline in 1990s to less than 1%, and eventually to 0.5% by 2007 (Wagner and Wellmer, 2008; KPMG, 2017; Damilek, 2017; NEITI, 2018; Olade, 2019). Currently, it is believed that the country's minerals and mining sector is under-achieving, and efforts are ongoing to revamp and energize the sector in the quest to diversify Nigeria's economy. The expected change will be transitional rather than transformational.

Nigeria is endowed with about 34 solid minerals identified in 450 locations in the country. Some of these minerals include gold, iron ore, cassiterite, columbite, wolframite, pyrochlore, monazite, marble, coal, limestone, clays, barites, lead-zinc, etc. and occur in the different metallogenic provinces of Nigeria. Out of which Government has classified seven of these minerals as strategic minerals, which include bitumen, gold, coal, iron ore, limestone, lead/zinc ores, limestone and barites. Other minerals with good economic prospects are: tin, tantalite, niobium, gypsum, gemstones, kaolin, etc. Less than 5% of these minerals are currently being mined, processed and marketed. These include coal, kaolin, barite, limestone, dolomite, gypsum, feldspar, gold, iron ore, lead-zinc, tin, and niobium and tantalum ores. The remaining 95% mineral ores, though in demand are untapped.

Presently, Solid minerals contribute as low as 3% to the country's GDP as against oil export which forms more than 95% of Nigeria's foreign exchange earnings. However, mineral exploration and production constitute significant parts of Nigeria's economy and remain keys to future economic growth. Despite a good fiscal regime for mining, the development of the mineral sector has been hampered by several constraints such as the dearth of proven mineral reserves, illegal mining, the lack of infrastructure, the lack of mineral production and processing capacity, the smuggling of minerals out of the country etc.

Mineral Resources Classification

1. Metallic Mineral Resources
2. Non-metallic Mineral Resources
3. Fuel Mineral Resources

1. Metallic mineral resources are minerals resources that contain metal in raw form, their appearances have metallic shine and they can be melted to obtain new products. Examples of metallic mineral resources include Gold, Silver, Copper, Tin, Iron, Lead, Zinc, Nickel, Chromium, and Aluminium.

2. Non-metallic mineral resources are minerals that do not contain extractable metals in their chemical composition; they include sand, stone, gravel, clay, gypsum halite, and Uranium.

3. Fuel mineral resources include coal, crude oil (petroleum) and natural gas

The South South Region

While the South South states are mainly associated with oil and gas production, the region is also endowed with several solid minerals. The vast reserves of non-renewable resources include granite, barites, marble, clay, gypsum, phosphate rock, feldspar, limestone sand and gravel. However, only a few of these minerals are currently being exploited. Previously significant foreign exchange earnings from coal, gold, tin, columbite, tantalite, lead, zinc and wolf Amite, were reduced dramatically with the onset of the oil boom and the neglect of the solid minerals in favour of oil production. The sector is generally neglected and a significant quantity of minerals and revenue are lost to illegal mining activities (Niger Delta Development Master Plan, 2011). In south south region, the solid minerals with the largest production are: industrial rocks; limestone, granite, laterite, sand, shale and clay that constitute over 95% of production by tonnage and 90% by value. The metallic minerals produced are mainly lead-zinc, manganese, gold, tin and columbite-tantalite which are exported and forming about 0.2% of total exports. The amount of gold produced is unknown, most of which is smuggled out of the country.

Efforts to diversify the Nigerian economy by increasing production and domestic utilization of industrial minerals, iron ores, bitumen and coal deposits form the core of a "roadmap" for the revitalization of the minerals and mining sector, with the main objective of increasing its contribution to 10% of GDP within the next decade. The absence of 'major' mining operations in the country is due to a lack of large or world-class mineral deposits which is attributed to the unfavourable geological and geotectonic environments. Above all, lack of reliable information on the spread and diversity of these natural resources is a challenge to development, since development is not just a 'governmental affair' with the right information, stakeholders, especially investors may be interested in joining efforts to develop the solid mineral sector.

Solid Minerals in South Southern Nigeria

Akwa Ibom	Clay, Lead/Zinc, Lignite, Limestone, Oil/Gas, Salt & Uranium
Bayelsa	Glay, Gypsium, Lead/Zinc, Lignite, Limestone, Maganese, Oil/Gas & Uranium
Cross River	Barite, Lead/Zinc, Lignite, Limestone, Manganese, Oil/Gas, Salt & Uranium
Delta	Clay, Glass-sand, Gypsium, Iron-ore, Kaolin, Lignite, Marble & Oil/Gas
Edo	Bitumen, Clay Dolomite, Phosphate, Glass-sand, Gold, Gypsium,Iron-ore, Lignite, Limestone, Marble & Oil/Gas
Rivers	Clay, Glass-Sand, Lignite, Marble & Oil/Gas

Source: http://www.nigeria.gov.ng

Geology is simply the study of the Earth. The various components of Geology include: the history of the earth as a planet since its formation; the materials of which the Earth is made; the structure of those materials; the processes (internal/external) acting upon the Earth, and the evolution of life forms. Earth materials such as, Human beings use Earth materials (water, air, metals and oil) every day and these materials are sourced from deep under the Earth. The tasks of the Geologists is to conduct studies that locate rocks that contain important metals (exploration), plan and employ the appropriate methods of removing the metals from the host rocks (exploitation). One of the important laws of modern Geology in the history of the Earth is the Principle of Uniformitarianism, which states that the "present is the key to the past and the future". Thus, the results of the processes that prevailed today are akin to those that prevailed in the past. Today we are concerned about population and climate change, food security, crimes and terrorism, among others. The geologists, environmentalists and other Scientists, policy makers, Economists and politicians therefore attempt to reconstruct past phenomena and how they have changed across time.

Geosciences

One of the roles of Geosciences is improving the quality and breadth of geological data gathered in a cost-efficient manner that will adequately drive investment growth in Nigeria. The dearth of reliable and bankable geological data on the available solid mineral resources in Nigeria discourages potential investors and poses real threat to the realization of the roadmap to solid minerals development. Given that mining is capital intensive, investors will typically be reluctant to expend resources unless there is reasonable basis to believe that the minerals present in a particular area exist in commercial quantities or that a combination of these minerals scattered in different locations can be realized in economical quantities.

System Analysis and Research Methods

This phase involves outlining the functions that the DW prototype can achieve, and an ideal working environment in which the DW prototype is delivered. Also in this stage, the decision maker's requirements are to understand the workings of users in relations to the Nigeria's and the public requirement and how they want to use the solution. What data are they currently making use of and what would they like to do with such data? The analysis stage involved a detailed learning of the existing system, leading to requirements of a new system. It also involved a detailed study of various operations performed by the system and their relationships. The requirements are gathered through a chain of interviews with the different stakeholders in the Solid Minerals sector. Answers from these users generate the requirements needed for further design of the DW prototype.

Functional requirements
Requirements are divided into functional requirements and non-functional requirements. Functional requirements are related with detailed functions, tasks or actions the system must support. The functional requirements are as follows:

i. The users should be able to access data from different applications in one single location.
ii. The users should be able to access data from different applications in a setup which is easy to operate.
iii. The system should be able to display figures and charts.
iv. The system should be able to update figures and charts.
v. The users should be able to download displayed figures and chats.
vi. The users should be able to print downloaded figures and charts.
vii. The system should be able to render reports in different formats which are useful to the users.
viii. The users should be able to access the system from anywhere in the world at any point in time (24/7).

Nonfunctional requirements
i. The system should be based on windows authentication so that user does not require to log on to the application many times but with a single sign-on.
ii. The front end application should be web enabled and no installation is required on users system.
iii. The front end application should be integrated into the existing portal like Share-Point and intranet.
iv. User level permission is required in order to protect the integrity of the data and restrict user's accessibility to data.
v. The system should perform very well at all times and should be easy to recover after system down time or shut down.
vi. The system should be able to keep up to-date information at all times.

Data Warehouse
A data warehouse (DW) is a pool of data produced to support decision making. It is also a repository of current and historical data of potential interest to managers. Data are usually structured to be available in a form ready-for-analytical processing activities (e.g. Online Analytical Processing (OLAP), data mining, querying, reporting and other decision supporting applications). A data warehouse is a subject-oriented, integrated, time-variant, non-volatile collection of data in support of management decision-making process. (Efraim *et al*, 2010). With a data warehouse and Online Analytical Processing (OLAP), users can perform better data analysis and gain better knowledge from the repository data. According to Stephen, Brobst and Joe Rarey (2003) Five (5) stages of decision support were identified in data-warehouse:

1. **Reporting**: This initial stage typically focuses on reporting from a single source of truth within the system. The biggest challenge in this stage is data integration.

2. **Analyzing**: In this stage, decision-makers focus less on what happened and more on why it happened. Analysis activities are concerned with drilling down beneath the numbers at a detailed level. Performance is more important in this stage in the data warehouse implementation process because the information repository is used much more interactively.

3. Predicting: As the system becomes well-entrenched in quantitative decision-making techniques and experiences the value proposition for understanding the "whats" and "whys" of its dynamics, the next step is to leverage information for predictive purposes. Advanced data mining methods often employ complex mathematical functions such as: logarithms, exponentiation, trigonometric functions and sophisticated statistical functions to obtain the predictive characteristics desired.

4. Operationalizing: Operationalization stage evolves into the realm of active data warehousing. Whereas stages 1 to 3 focus on strategic decision-making within the system, this stage focuses on tactical decision support. Operationalizing typically means providing access to information for immediate decision-making in the field. Two examples are: 1) inventory management and control with just-in-time replenishment and 2) scheduling and routing for package delivery.

5. Active Warehousing: An active data warehouse delivers information and enables decision support throughout the system rather than being confined to strategic decision-making processes. However, tactical decision support does not replace strategic decision support; rather, an active data warehouse supports the coexistence of both types of workloads.

Data warehousing is an electronic method of organizing information. The responsibility of enterprise data warehousing is to provide data repositories that support the organizational activities that data governance and information management have identified as necessary.

Why it matters
A data warehouse essentially combines information from several sources into one comprehensive database. For example, in the business world, a data warehouse might incorporate customer information from a company's point-of-sale systems (the cash registers), its website, its mailing lists and its comment cards. Alternatively, it might incorporate all the information about employees, including time cards, demographic data, salary information, etc. By combining all of this information in one place, a company can analyze its customers in a more holistic way, ensuring that it has considered all the information available. Data warehousing also makes data mining possible, which is the task of looking for patterns in the data that could lead to higher sales and profits. There are different ways to establish a data warehouse and many pieces of software that help different systems "upload" their data to a data warehouse for analysis. However, the basic idea is to first extract data from all the individual sources (cash registers, time clocks, office computers), remove redundant data and organize the data into a consistent format that can be queried. Government, organizations and companies with data warehouses can have an advantage in resource identification, development, marketing, pricing strategy, production time, historical analysis, forecasting and customer and investor satisfaction. However, data warehouses also can be very expensive to design and implement, and sometimes their construction makes them unwieldy.

How data mining works
While large-scale information technology has been evolving separate transaction and analytical systems, data mining provides the link between the two. Data mining software analyzes relationships and patterns in stored transaction data based on open-ended user queries. Several types of analytical software are available: statistical, machine learning, and neural networks. Generally, any of four types of relationships are sought in the data warehousing technology:

1. **Classes**: Stored data is used to locate data in predetermined groups. For example, a solid mineral exploitation chain could mine solid mineral data to determine location, characteristics and what they are typically used for. This information could be used to increase traffic for the resource.

2. **Clusters**: Data items are grouped according to logical relationships or consumer/user preferences. For example, data can be mined to identify types of solid mineral or investor/processor affinities.

3. **Associations**: Data can be mined to identify associations. The **beer-diaper** example is an example of associative mining.

4. **Sequential patterns**: Data is mined to anticipate behaviour patterns and trends. For example, an outdoor equipment retailer could predict the likelihood of a backpack being purchased based on a consumer's purchase of sleeping bags and hiking shoes.

Data mining consists of five major elements:

- Extracting, transforming, and loading transaction data onto the data warehouse system.
- Storing and managing the data in a multidimensional database system.
- Providing data access to geo scientists, business analysts and information technology professsionals.
- Analyzing the data by application soft-ware.
- Presenting the data in a useful format, such as a graph, charts or tables.
- Different levels of analysis are available:

1. Artificial neural networks: Non-linear predictive models that learn through training and resemble biological neural networks in structure.

Genetic algorithms: Optimization techniques that use processes such as: genetic combination, mutation, and natural selection in a design based on the concepts of natural evolution.

Decision trees: Tree-shaped structures that represent sets of decisions. These decisions generate rules for the classification of a dataset.

Specific decision tree methods include: Classification and Regression Trees (CART) and Chi Square Automatic Interaction Detection (CHAID). CART and CHAID are decision tree techniques used for classification of a dataset for a data warehouse system. They provide a set of rules that can be applied to a new (unclassified) dataset to predict which records will have a given outcome. CART segments a dataset by creating 2-way splits while CHAID segments using chi square tests to create multi-way splits. CART typically requires less data preparation than CHAID (Brobst and Rarey, 2003).

2. Nearest neighbour method: A technique that classifies each record in a dataset based on a combination of the classes of the k record(s) most similar to it in a historical dataset (where k 1). Some-times called the k-nearest neighbour technique.

3. Rule induction: The extraction of useful if-then rules from data based on statistical significance.

4. Data visualization: The visual interpretation of complex relationships in multi-dimensional data. Graphics tools are used to illustrate data relationships in this case (Palace 1996).

The Motivation for the Data Warehouse system

The day-to-day operations of the school rely heavily upon the OLTP system. The managers make use of OLTP system for management information system. The need for strategic decision necessitates the development of the data warehouse. The OLTP can only handle few historical data, which are not enough to make strategic decisions. With a data warehouse and OLAP, database manager can perform roll-up and drill-down operations to **Solid Minerals registered** by type, by class, by name, by location, by usefulness or any combination desired. The system will provide decision support that is flexible and user-friendly. Policy makers (Ministry of Solid Minerals, Geoscientists) would like to use the data-warehouse to answer questions such as: What is the quantity of Uranium found in the South South Region? At which state or location is it found? Wtaht can it be used for? Can it be exploited and processed or can it be exported in crude form? What is the cost of processing this mineral in Nigeria, etc. These types of questions need a lot of historical data to generate which the current system the OLTP system cannot support

Designing the Data Ware-House

The following discussion outlines the process of the data warehouse design. It involves the logical design, the OLAP design, and Data mining design.

This image has been removed due to copyright issues.

Source: The DW Lifecycle (DWLC) Model (WhereScape, 2013)

The logical Design
The logical design of data-warehouse is defined by the dimensional data modelling approach. The dimensioning design process followed in this project adheres to the methodology described by Kimball and Ross (2002). Unlike the Entity Relationship (ER) and Unified Modeling Language (UML) data modeling processes, the logical design of Data Warehouse (DW) is defined by the dimensional data modelling approach. To minimize the join operations which slow down queries, normalization is not the guiding principle in DW design. A schema is a collection of database objects, including tables, views, indexes, and synonyms. There is a variety of ways of arranging schema objects in the schema models designed for data warehousing. The following are the two types of schemas commonly used in dimensional data modelling.

Star schema: This is perhaps the simplest data warehouse schema. It is called a star schema because the entity-relationship diagram of this schema resembles a star, with points radiating from a central table. The centre of the star consists of a large fact table and the points of the star are the dimension tables. A star schema is characterized by one or more very large fact tables that contain the primary information in the data warehouse, and a number of much smaller dimension tables (or lookup tables), each of which contains information about the entries for a particular attribute in the fact table. Star schema facilitates quick response to queries. The core detailed values are stored in the fact table. The dimensional information and hierarchies are kept in the dimension tables (Akintola, Adetunmbi and Adeola, 2011).

Snowflake schema: The snowflake schema is a more complex data warehouse model than a star schema, and is a type of star schema. It is called a snowflake schema because the diagram of the schema resembles a snowflake. Snowflake schemas normalize dimensions to eliminate redundancy. That is, the dimension data has been grouped into multiple tables instead of one large table. For example, a product dimension table in a star schema might be normalized into a products table, a product category table, and a product manufacturer table in a snowflake schema. While this saves space, it increases the number of dimension tables and requires more foreign key joins. The result is more complex queries and reduced query performance

Implementation

Data WarehouseTransporting: Data from OLTP Database to Data Warehouse
SQL Server provides the tool SSIS to assist in transporting data in and out of a database.It involves creation of the dimension table's data structure, followed by the creation of integration service project using the Business Intelligence package in the SQL server. After creating the project, a data-source is then created followed by the creation of OLEDB data-source and destination.

DW Prototyping

User requirements
i. Ability to use data from DW Prototype database for data mining activities.
ii. Ability to perform complex queries to produce reports and analysis with little effort.
iii. Ability to download view and print information from the DW prototype and use it for further analysis.
iv. System reliability at all times.

System requirements
For the purpose of this study, the system requirements are divided into Hardware and Software componnents.

The Hardware system requirements are:

i. Central Processing Unit (CPU): Intel (R) CORE i5-2430M CPU @ 2.4GHz.
ii. Installed memory (RAM): 4.00 GB.
iii. Monitor: Super VGA (AGP 32MB).
iv. Peripherals: Keyboard, Mouse, USB and etc.

The Software system requirements are:

i. System type: 64-bit Operating System (Windows 7 Ultimate).
ii. Database engine: Microsoft SQL Server 2005 and 2012.
iii. ETL tools: SQL Server Integration Service (SSIS).
iv. Visual Studio 2010 Professional
v. Programming Language: ASP .NET, XML, CSS and C#.
vi. Web browser: Goggle Chrome, Firefox, Opera and Internet Explorer.

System Design
This phase is focused to translating the systems requirements into a set of specifications through physical data models or Data Marts for the DW prototype. Inmon ((1992, 1993) and Kimball (2002) have different approaches to the design of the DW. For the purpose of this study, the Kimball's design approach is adopted. The data of Solid Minerals from each state in the South South region is treated as a Data Mart and it is designed from start to finish. One of the main aims of the DW is to extract data from different OLTP or flat file sources and combine them in a single source for easy access and make best use of the data. The two processes of DW are data load and access. The design of the database starts with the principle and theories of database design and the rule that support the stakeholders' needs.

The Software Development Methodology Adopted
In a DW development, the DW life cycle and the Software engineering development methodologies can contribute meaningfully because it creates a well-established approach and technique for the development processes. DW development is an iterative process as such the "Realistic Waterfall Model" is nused for the prototype development. The Realistic Waterfall model is a risk-oriented iterative enhancement, and recognizes that requirements are not always available and clear when the system is first implemented. Since designing and building a DW is an iterative process, the realistic waterfall model is one of the development methodologies of choice (Inuwa and Oye, 2016). This is to ensure that any business

requirements not clear at the beginning of the analysis stage can often be revisited. The Figure 2 shows a realistic waterfall model in the spiral model fashion of the DW life cycle.

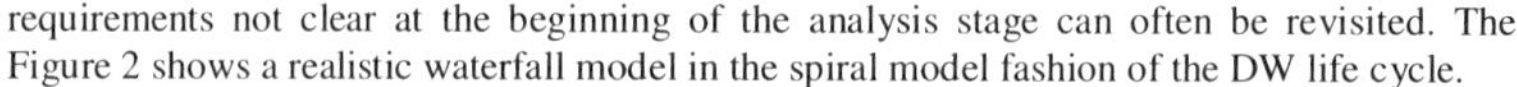

This image has been removed due to copyright issues.

A Realistic Waterfall Model in the Spiral Model of the DW Life Cycle. Adapted from: (Inuwa and Oye,(2016)

The researchers started with Data Mart design by identifying the measures of the fact tables. The measures are the basis for feedback information that the decision makers require. For the purpose of this study, we adopted the star schema for the prototype DW database design.

Results and Implementation

System Prototype Development and Validation

The system prototype development is the actual application of the analysis and design that has been carried out. In this phase of the study, the DW (Fact and dimension tables) prototype, the ETL (Extract, Transform and Load) and the front end application were designed. Validation process involves the confirmation by examination that an information system (DW prototype) has been realized appropriately and it is in conformity with the User's needs and intended use. The researchers have designed the actual physical Data Mart Models of all the independent solid mineral Data Marts on an SQL Server Database Management System 2005.

Creating and loading of the DW prototype database

The Solid Minerals DW database was created on an SQL Server 2012, and the data loading was also done through the SSIS package. Fact and Dimension tables were pushed into the DW Prototype database of the DW Prototype Model by Star Schema. The DW Prototype database is the one that integrates the Fact and Dimension tables of solid minerals into the SQL Server 2012. It is part of the SQL Server application that was used for the database repository. The DW Prototype database was populated with the correct data of good quality that so that the data repository can be used. DSS is all about making the best use of the available data in order to make better decision about the resources through the DW platform.

System Verification, Sample Reports and Analysis
The best way to verifying the simulated data in the DW Prototype is to prepare queries on
these data. In this study, certain sample reports were presented and the purpose of these
reports is to demonstrate the usefulness of the DW approach to decision making as far as
solid minerals are concerned. Even though these reports were based on the user requirements,
it was not planned to design a full set of management reports and analysis. This set of sample
reports can be used as a basis for generating more comprehensive sets by applying complex
queries on the data. The reports generated from this DW Prototype Model have been
classified based on State, Location, Solid Mineral, Type, quantity and Uses.

The User Interface (Input and Output designs)
The Interface design comprises of both input and output designs. The SQL Server reporting
Services 2012 is used to build reports based on the data contained in DW prototype database.
The SQL Server Reporting Services 2012 is a special Server that is used for reporting and
analytical service; it is linked with the DW database Server for reporting and analysis
activities. The system input was designed so as to receive data from the users and
administrators of the DW system. A system output is the most important component of a
working system, because the interactivity of the system depends on its output. This is the
main reason why the output of an information system determines the effectiveness and
efficiency of the system. The system output present information to the users and
administrators based on the required queries performed on the DW database. Sample reports
are produced and presented and the motive of these reports is to demonstrate the usefulness
of the DW approach. The reports were interactively produced by the use of the SQL Server
Reporting tool based on user's requirements. The reports were scheduled based on States and
Soild minerals reports.

Conclusion
Given the over dependence of the Nigerian economy on volatile global of oil prices, coupled
with the neglect of other sectors that had hitherto sustained the Nigerian economy, it has now
become imperative for the Nigerian economy to be diversified. Predicated on this scenario
(volatility of oil prices), the economy will be subjected to fiscal shocks with attendant
problems such as fiscal deficits, like what Nigerians experienced this year 2019. Thus, the
ministry of solid minerals main policy objective which is the increase mineral production and
value addition by focusing on exploration, exploitation, processing utilization and marketing
of mineral resources of the country needs to be strengthened to achieve these policy
objectives. In this paper, we have been able to demonstrate the process of designing and
developing data-warehouse and data mining applications using SQL server business
intelligence Development tools using for soild minerals identified in the South South States
of Nigeria. It is noteworthy, however that this technique can be applied to all states of the
Federation including the Federal Capital Territory or any organization wishing to implement
business intelligence as part of their strategic decision support operations. The power of Data-
Warehousing in data analysis is tremendous and data-mining can discover hidden treasures
not only in the data-warehouse, but in other parts of the country, which thorough
investigation andn research on solid mineral deposits has not yet been performed.
Organizations, particularly in Nigeria can begin to implement this project as part of their
strategic decision making process tools to aid Government in major economis decisions and
timely actions. With it they can begin to see trends of things historically as their day to day
operational data accumulates over the years. They can forecast the future using neural
Networks, regression analysis, and other data mining operations incorporated into SQL server

Business Intelligence. Data-warehouse for Solid Mineral resources is the solution we need at the moment to catapult this vibrant, but dull economic sector to the next level for industrialization and economic development.

REFERENCES

Akintola K.G., Adetunmbi A.O. and Adeola O.S. (2011) Building Data Warehousing and Data Mining arom Course Management Systems: A Case Study of Federal University of Technology (FUTA) Course Management Information Systems. Information Technology for People-Centred Development (ITePED 2011). Nigeria Computer Society (NCS): 10 th international Conference –July 25-29,

Bamali, U. S. (2011) A review of Nigerian metallic minerals for technological development. *Natural Resources*, v. 2, p. 87-91.

Bauwens , Christian (2005) "Oracle Database Sample Schemas 10g Release 2 (10.2)." Oracle. June 2005. Oracle Corporation. 27 May 2011. http://download.oracle.com/docs

Damilek, D.D. (2017) Nigeria solid mineral resource potential: An overview. Paper in Researchgate, 17pp.

DeGrace, P. & Stahl, L. H. (1990). *Wicked Problems, Righteous Solutions: A Catalogue of Modern Software Engineering Paradigms.* Reprinted with permission of Prentice Hall, Englewood Cliffs, New Jersey, pp. 116, 117, and 127.

Demarest, M. (2013). Data Warehouse Prototyping: Reducing Risk, Securing Commitment and Improving Project Governance. Retrieved June 28, 2014, from: http// www.wherescape.com

Department of Planning, Research and Statistics (DPRS) Ministry of Mines and Steel Development (MMSD) (2017) Nigeria's Mining Sector Investment Promotion Brochure October

Efraim T., Jay E., Teng-Peng L., Ramesh S, (2010). *Decision Support and Business Intelligence Systems* (8th ed) Prentice Hall.Infogold Data Warehouse, Data Mart, Data Mining, and Decision Support Resources,http://infogoal.com/dmc/dmcdwh.htm

Fayemi, Kayode and Onwurah, Chi (2016) Nigeria's Solid Minerals Sector: Alternative Investment Opportunities. All Party Parliamentary Group on Nigeria Transcript www.chathamhouse.org

Hughes, Ralph (2013) Estimating and Segmenting Projects in Agile Data Warehousing Project Management, Data Warehousing Project. www.sciencedirect.com

Hughes, Ralph (2016) Eliminating Risk through Nested Iterations, in Agile Data Warehousing for the Enterprise, Data Warehousing Project. www.sciencedirect.com

Inmon, W. H. (1992. a). *Building the Data Warehouse*. New York. John Wiley & Sons, Inc.

Inmon, W.H., (1993). *Building the Data Warehouse.* A Wiley QED publication. John Wiley and Sons, Inc. New York 123-133.

Inuwa, Ibrahim and Oye, N. D. (2016) Modelling and Simulation of a Decision Support System Prototype Built on an Improved Data Warehousing Architecture for the School of Postgraduate, MAUTECH, Yola – Nigeria *Information and Knowledge Management* Vol.6, No.1, 2016 www.iiste.org ISSN 2224-5758 (Paper) ISSN 2224-896X (Online)

Kimball R. and Ross M. (2002). *The Data Warehouse Toolkit,* second edition. John Wiley and Sons, Inc., USA.

Kimball, R. and Ross, M. (2002). *The Data Warehouse Toolkit: The Complete Guide to Dimensional Modeling.* New York. John Wiley & Sons, Inc.

KPMG (2017) Nigerian Mining Sector Brief 2017. Publication of KPMG Nigeria. 20pp.

MMSD (2016) Roadmap for the growth and development of the Nigerian mining industry, 99pp.

MMSD (2017) Nigeria's Mining and Metals Sector. Investment promotion brochure. Ministry of Mines and Steel Development. 28pp

NEITI (2017) 2016 Annual Progress Report, 62pp.

NEITI (2018) Nigeria EITI 2017 Annual Progress Report. 79pp.

Niger Delta Development Commission (2011) Niger Delta Development Master Plan, 2011

Olade, M. A. (2019) Solid Mineral Deposits and Mining in Nigeria: - A Sector in Transitional Change *Achievers Journal of Scientific Research* Volume 2, Issue 1, p. 1-16 ISSN 2504-0141 www.researchgate.net

Olade, M.A. (2018) Economic Geology and Mineral Deposits of Nigeria. Prescott Resource Publishers, 505 pp.

Pace, Bill (1996). Data Mining Technology Note prepared for Management 274A Anderson Graduate School of Mana-gement at UCLA http://www.ander-son.ucla.edu/

Stephen Brobst, Joe Rarey (2003). Five Stages of Data Warehouse Decision Support Evolution http://dssre-sources.com/papers

USGS (2017) 2014 Minerals Yearbook, Nigeria. United States Geological Survey, 8 pp.

Wagner, M and Wellmer, F.W. (2008) Global mineral resources: occurrence and distribution. In Environmental and Engineering Geology, v. 3, UNESCO-EOLSS Report, 26pp.

Where Scape, (2013). Data Warehouse Prototyping. Retrieved June 28, 2014, from: http//www.wherescape.com

www.nigeria.gov.ng

www.nigeria.gov.ng/index.php/2016-04-06-08-38-30/nigeria-natural-resources

YOUR KNOWLEDGE HAS VALUE

- We will publish your bachelor's and
  master's thesis, essays and papers

- Your own eBook and book -
  sold worldwide in all relevant shops

- Earn money with each sale

Upload your text at www.GRIN.com
and publish for free